# TRACING NUMBER

## PRESCHOOLERS PRACTICE WRITING WORKBOOK, KIDS AGES 3-5

### Book 13

https://tinyurl.com/tracingletterforkidsages3-5

# COPYRIGHT NOTICE

Copyright © 2017 by **TRACING NUMBER PRESCHOOLERS PRACTICE WRITING WORKBOOK, KIDS AGES 3-5**.

All rights reserved. This book or any portion thereof may not be reproduced or used in any manner whatsoever without the express written permission of the publisher except for the use of brief quotations in a book review.

# CONTENTS

| | |
|---|---|
| **INTRODUCTION** | 1 |
| **HOW TO USE THIS BOOK** | 2 |
| *NUMBERS 1-50 EXERCISE 1* | 3 |
| *NUMBERS 1-50 EXERCISE 2* | 4 |
| *NUMBERS 1-50 EXERCISE 3* | 5 |
| *NUMBERS 1-50 EXERCISE 4* | 6 |
| *NUMBERS 1-50 EXERCISE 5* | 7 |
| *NUMBERS 1-50 EXERCISE 6* | 8 |
| *NUMBERS 1-50 EXERCISE 7* | 9 |
| *NUMBERS 1-50 EXERCISE 8* | 10 |
| *NUMBERS 1-50 EXERCISE 9* | 11 |
| *NUMBERS 1-50 EXERCISE 10* | 12 |
| *NUMBERS 1-50 EXERCISE 11* | 13 |
| *NUMBERS 1-50 EXERCISE 12* | 14 |
| *NUMBERS 1-50 EXERCISE 13* | 15 |
| *NUMBERS 1-50 EXERCISE 14* | 16 |
| *NUMBERS 1-50 EXERCISE 15* | 17 |
| *NUMBERS 1-50 EXERCISE 16* | 18 |
| *NUMBERS 1-50 EXERCISE 17* | 19 |
| *NUMBERS 1-50 EXERCISE 18* | 20 |
| *NUMBERS 1-50 EXERCISE 19* | 21 |
| *NUMBERS 1-50 EXERCISE 20* | 22 |
| *NUMBERS 1-50 EXERCISE 21* | 23 |
| *NUMBERS 1-50 EXERCISE 22* | 24 |
| *NUMBERS 1-50 EXERCISE 23* | 25 |
| *NUMBERS 1-50 EXERCISE 24* | 26 |
| *NUMBERS 1-50 EXERCISE 25* | 27 |
| *NUMBERS 1-50 EXERCISE 26* | 28 |
| *NUMBERS 1-50 EXERCISE 27* | 29 |
| *NUMBERS 1-50 EXERCISE 28* | 30 |
| *NUMBERS 1-50 EXERCISE 29* | 31 |
| *NUMBERS 1-50 EXERCISE 30* | 32 |
| **YOU MIGHT ALSO BE INTERESTED IN MY NEWEST COLLECTION OF BOOKS** | 35 |

# INTRODUCTION

A child who learns to trace NUMBERS at home, at the early age with their loving parent or caregiver, grows in self-confidence and independence. This promotes greater maturity, increases discipline and lays the basis for moral literacy. A child who begins with early learning books has a distinct advantage over his or her peers. One of the big advantages being there is no psychological pressure. The reason why parent understand why early learning is important.

# HOW TO USE THIS BOOK

Practice, Practice, Practice makes life easier and worthwhile so train your child analytical mind by tracing letters and numbers in the alphabet the conventional way through handwritings as the saying said: "**Young children need writing to help them learn about reading, they need reading to help them learn about writing; and they need oral language to help them learn about both.**"

**\*\*\*Make your tracing at your own style and convenience all individuals has its own uniqueness.\*\*\*\***

# NUMBERS 1-50 EXERCISE 1

| | |
|---|---|
| 33 | Thirty three |
| 21 | Twenty one |
| 23 | Twenty three |
| 31 | Thirty one |
| 36 | Thirty six |
| 17 | Seventeen |
| 30 | Thirty |
| 28 | Twenty eight |
| 18 | Eighteen |
| 6 | Six |

# NUMBERS 1-50 EXERCISE 2

| | |
|---|---|
| 48 | Forty eight |
| 3 | Three |
| 17 | Seventeen |
| 46 | Forty six |
| 22 | Twenty two |
| 38 | Thirty eight |
| 5 | Five |
| 14 | Fourteen |
| 31 | Thirty one |
| 33 | Thirty three |

# NUMBERS 1-50 EXERCISE 3

| | |
|---|---|
| 34 | Thirty four |
| 36 | Thirty six |
| 44 | Forty four |
| 38 | Thirty eight |
| 26 | Twenty six |
| 20 | Twenty |
| 27 | Twenty seven |
| 23 | Twenty three |
| 16 | Sixteen |
| 43 | Forty three |

# NUMBERS 1-50 EXERCISE 4

| | |
|---|---|
| 24 | Twenty four |
| 10 | Ten |
| 22 | Twenty two |
| 36 | Thirty six |
| 17 | Seventeen |
| 41 | Forty one |
| 5 | Five |
| 16 | Sixteen |
| 43 | Forty three |
| 33 | Thirty three |

# NUMBERS 1-50 EXERCISE 5

26 — Twenty six
19 — Nineteen
32 — Thirty two
34 — Thirty four
42 — Forty two
11 — Eleven
17 — Seventeen
50 — Fifty
46 — Forty six
31 — Thirty one

# NUMBERS 1-50 EXERCISE 6

| | |
|---|---|
| 12 | Twelve |
| 24 | Twenty four |
| 35 | Thirty five |
| 47 | Forty seven |
| 10 | Ten |
| 3 | Three |
| 1 | One |
| 9 | Nine |
| 45 | Forty five |
| 39 | Thirty nine |

# NUMBERS 1-50 EXERCISE 7

39 — Thirty nine
42 — Forty two
25 — Twenty five
30 — Thirty
45 — Forty five
22 — Twenty two
50 — Fifty
24 — Twenty four
15 — Fifteen
43 — Forty three

# NUMBERS 1-50 EXERCISE 8

8     Eight

32    Thirty two

50    Fifty

7     Seven

34    Thirty four

18    Eighteen

43    Forty three

37    Thirty seven

6     Six

19    Nineteen

# NUMBERS 1-50 EXERCISE 9

36 Thirty six
9 Nine
48 Forty eight
35 Thirty five
44 Forty four
2 Two
31 Thirty one
10 Ten
50 Fifty
25 Twenty five

# NUMBERS 1-50 EXERCISE 10

10 Ten
5 Five
33 Thirty three
12 Twelve
49 Forty nine
34 Thirty four
27 Twenty seven
1 One
32 Thirty two
2 Two

# NUMBERS 1-50 EXERCISE 11

| | |
|---|---|
| 39 | Thirty nine |
| 5 | Five |
| 1 | One |
| 36 | Thirty six |
| 40 | Forty |
| 33 | Thirty three |
| 31 | Thirty one |
| 18 | Eighteen |
| 46 | Forty six |
| 34 | Thirty four |

# NUMBERS 1-50 EXERCISE 12

| | |
|---|---|
| 1 | One |
| 4 | Four |
| 16 | Sixteen |
| 17 | Seventeen |
| 13 | Thirteen |
| 46 | Forty six |
| 15 | Fifteen |
| 5 | Five |
| 20 | Twenty |
| 34 | Thirty four |

# NUMBERS 1-50 EXERCISE 13

5     Five

4     Four

14    Fourteen

49    Forty nine

3     Three

23    Twenty three

39    Thirty nine

47    Forty seven

45    Forty five

17    Seventeen

# NUMBERS 1-50 EXERCISE 14

19 Nineteen
10 Ten
45 Forty five
46 Forty six
22 Twenty two
12 Twelve
13 Thirteen
35 Thirty five
38 Thirty eight
1 One

# NUMBERS 1-50 EXERCISE 15

38 Thirty eight
3 Three
27 Twenty seven
19 Nineteen
35 Thirty five
12 Twelve
20 Twenty
49 Forty nine
37 Thirty seven
44 Forty four

# NUMBERS 1-50 EXERCISE 16

| | |
|---|---|
| 1 | One |
| 50 | Fifty |
| 38 | Thirty eight |
| 5 | Five |
| 16 | Sixteen |
| 21 | Twenty one |
| 23 | Twenty three |
| 14 | Fourteen |
| 34 | Thirty four |
| 9 | Nine |

# NUMBERS 1-50 EXERCISE 17

8     Eight

18    Eighteen

36    Thirty six

12    Twelve

26    Twenty six

5     Five

13    Thirteen

27    Twenty seven

50    Fifty

35    Thirty five

# NUMBERS 1-50 EXERCISE 18

28 Twenty eight
33 Thirty three
19 Nineteen
22 Twenty two
46 Forty six
21 Twenty one
43 Forty three
6 Six
27 Twenty seven
34 Thirty four

# NUMBERS 1-50 EXERCISE 19

13 Thirteen
18 Eighteen
10 Ten
1 One
19 Nineteen
6 Six
46 Forty six
4 Four
38 Thirty eight
22 Twenty two

# NUMBERS 1-50 EXERCISE 20

| | |
|---|---|
| 11 | Eleven |
| 34 | Thirty four |
| 49 | Forty nine |
| 14 | Fourteen |
| 12 | Twelve |
| 26 | Twenty six |
| 8 | Eight |
| 50 | Fifty |
| 29 | Twenty nine |
| 5 | Five |

# NUMBERS 1-50 EXERCISE 21

41 Forty one
33 Thirty three
2 Two
3 Three
38 Thirty eight
47 Forty seven
14 Fourteen
25 Twenty five
13 Thirteen
10 Ten

# NUMBERS 1-50 EXERCISE 22

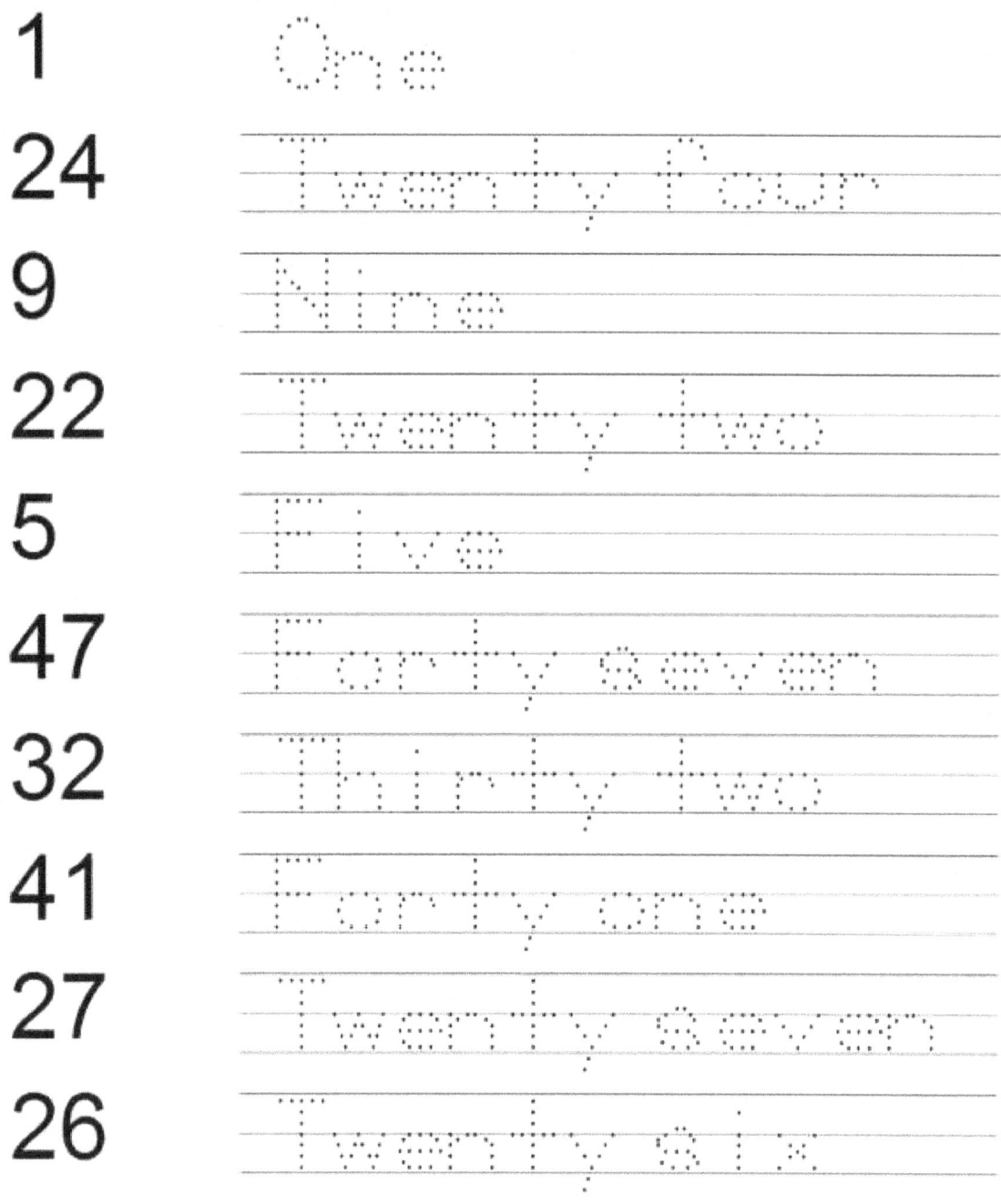

1 One
24 Twenty four
9 Nine
22 Twenty two
5 Five
47 Forty seven
32 Thirty two
41 Forty one
27 Twenty seven
26 Twenty six

# NUMBERS 1-50 EXERCISE 23

43 Forty three
17 Seventeen
45 Forty five
10 Ten
24 Twenty four
28 Twenty eight
13 Thirteen
16 Sixteen
42 Forty two
30 Thirty

# NUMBERS 1-50 EXERCISE 24

| | |
|---|---|
| 29 | Twenty nine |
| 15 | Fifteen |
| 47 | Forty seven |
| 37 | Thirty seven |
| 26 | Twenty six |
| 16 | Sixteen |
| 19 | Nineteen |
| 43 | Forty three |
| 20 | Twenty |
| 22 | Twenty two |

# NUMBERS 1-50 EXERCISE 25

| | |
|---|---|
| 25 | Twenty five |
| 11 | Eleven |
| 15 | Fifteen |
| 23 | Twenty three |
| 13 | Thirteen |
| 38 | Thirty eight |
| 46 | Forty six |
| 12 | Twelve |
| 45 | Forty five |
| 1 | One |

# NUMBERS 1-50 EXERCISE 26

| | |
|---|---|
| 16 | Sixteen |
| 45 | Forty five |
| 29 | Twenty nine |
| 22 | Twenty two |
| 32 | Thirty two |
| 24 | Twenty four |
| 26 | Twenty six |
| 44 | Forty four |
| 39 | Thirty nine |
| 33 | Thirty three |

# NUMBERS 1-50 EXERCISE 27

| | |
|---|---|
| 12 | Twelve |
| 33 | Thirty three |
| 11 | Eleven |
| 35 | Thirty five |
| 26 | Twenty six |
| 16 | Sixteen |
| 15 | Fifteen |
| 30 | Thirty |
| 28 | Twenty eight |
| 4 | Four |

# NUMBERS 1-50 EXERCISE 28

32  Thirty two
39  Thirty nine
15  Fifteen
7   Seven
46  Forty six
16  Sixteen
42  Forty two
19  Nineteen
48  Forty eight
14  Fourteen

# NUMBERS 1-50 EXERCISE 29

| | |
|---|---|
| 49 | Forty nine |
| 25 | Twenty five |
| 35 | Thirty five |
| 21 | Twenty one |
| 10 | Ten |
| 20 | Twenty |
| 26 | Twenty six |
| 38 | Thirty eight |
| 39 | Thirty nine |
| 16 | Sixteen |

# NUMBERS 1-50 EXERCISE 30

22  Twenty two
1   One
34  Thirty four
4   Four
20  Twenty
36  Thirty six
2   Two
37  Thirty seven
19  Nineteen
11  Eleven

TO SAY THANK YOU FOR PURCHASING THIS BOOK I AM GIVING AWAY A FREE COPY OF MY LATEST CHILDREN'S BOOK OR SUDOKU BOOK!

JUST GO TO THE LINK BELOW TO AND COMMENT TO RECEIVED YOUR COPY

http://sudokuprintable.blogspot.com

THIS IS A GOOD EXAMPLE FOR THOSE OF YOU WITH A SUBSCRIBER LIST

YOU MIGHT ALSO BE INTERESTED IN MY NEWEST COLLECTION OF BOOKS

CHECK OUT THE BOOKS ON THE NEXT PAGE

# TIMES ULTIMATE MIND GAMES BOOK KILLER SUDOKU PUZZLE BOOK 1

**FIND OUT MORE HERE:**

https://tinyurl.com/sudoku-9210

# 500 SUDOKU PUZZLE BOOK WITH ANSWERS

https://tinyurl.com/sudoku501

# TRACING LETTER PRESCHOOLERS PRACTICEWRITING ABC ALPHABET WORKBOOK,KIDS AGES 3-5 BOOK 1 : BY BRIGHTER HAND

https://tinyurl.com/tracingletterforkidsages3-5

www.ingramcontent.com/pod-product-compliance
Lightning Source LLC
Chambersburg PA
CBHW082222220526
45470CB00010B/3269